Tina Skinner # Pavers 101

4880 Lower Valley Road, Atglen, Pa 19310

Patios and Other Projects You Can Do

Schiffer Books are available at special discounts for bulk purchases for sales promotions or premiums. Special editions, including personalized covers, corporate imprints, and excerpts can be created in large quantities for special needs. For more information contact the publisher:

Published by Schiffer Publishing Ltd.
4880 Lower Valley Road
Atglen, PA 19310
Phone: (610) 593-1777; Fax: (610) 593-2002
E-mail: Info@schifferbooks.com

Please visit our web site catalog at www.schifferbooks.com

We are always looking for people to write books on new and related subjects. If you have an idea for a book, please contact us at the above address.

This book may be purchased from the publisher.
Include $5.00 for shipping. Please try your bookstore first.
You may write for a free catalog.

In Europe, Schiffer books are distributed by:
Bushwood Books
6 Marksbury Ave.
Kew Gardens
Surrey TW9 4JF
England
Phone: 44 (0)208 392-8585
Fax: 44 (0)208 392-9876
E-mail: Info@bushwoodbooks.co.uk
Website: www.bushwoodbooks.co.uk
Free postage in the UK. Europe: air mail at cost.
Try your bookstore first.

Copyright © 2008 by Schiffer Publishing, LTD.
Library of Congress Control Number: 2001012345

All rights reserved. No part of this work may be reproduced or used in any form or by any means—graphic, electronic, or mechanical, including photocopying or information storage and retrieval systems—without written permission from the publisher.
The scanning, uploading and distribution of this book or any part thereof via the Internet or via any other means without the permission of the publisher is illegal and punishable by law. Please purchase only authorized editions and do not participate in or encourage the electronic piracy of copyrighted materials.
"Schiffer," "Schiffer Publishing Ltd. & Design," and the "Design of pen and ink well" are registered trademarks of Schiffer Publishing Ltd.

Designed by RoS
Type set in Zurich BLK BT/NewBskvll BT

ISBN: 978-0-7643-3053-7

Printed in China

Contents

Introduction 5

A Walkway to Remember 6
  Laying the Groundwork
  Basketweave Pattern
  Herringbone Pattern
  Running Bond Pattern
  Finishing the Walkway
  Gallery of Patterns
  Paver Patterns at Work

Patio Project with Fire Pit 32
  Calculating Your Needs
  A Study in Similar Masterpieces

Putting Up Walls 68
  Walls to Remember

Permeable Driveway Project 78

Gallery of Great Ideas 88
  Driveways
  Entryway Patios
  Perfect Paths
  Wonderful Walls
  Steps
  More Pretty Patios

Foreword

Homeowners are expanding their living spaces. But instead of building an addition or finishing a basement room, they're growing in the other direction with outdoor kitchens, entertainment areas, firepits, and spas.

This new outdoor living movement has fueled explosive growth in interlocking concrete paving stones—the preferred pavement of do-it-yourselfers—and as a result, there are more choices than ever from which to build the landscape of your dreams.

Interlocking concrete paving stones—or pavers—will breathe new life into your outdoor environment with an exciting array of colors, textures, and patterns to complement any patio, courtyard, walkway, or drive. Say goodbye to drab concrete and asphalt, and say hello to the elegant look and inviting warmth of pavers.

Durability of concrete pavers is second to none. Paver surfaces have all the strength of solid pavements, yet they remain flexible and resistant to cracking or failure throughout the freeze-thaw cycles of the seasons. Properly installed, your paver surface will provide many years of virtually maintenance-free enjoyment.

Inside this book, you'll find everything you need to know to install your own paver patio, walkway or driveway, along with hundreds of design ideas to match every skill level. You'll also get a look at the green landscaping solution of the future—permeable pavers—and how they can help beautify your yard and protect the environment at the same time. If you have further questions about interlocking concrete pavers, I invite you to visit our web site, www.willowcreekpavingstones.com.

So take a walk through these pages, and you'll soon be on the path to a new way of outdoor living!

Todd Strand
President
Willow Creek Concrete Products
Oakdale, Minnesota

Acknowledgements

The following people contributed to this book:

Project Coordinator: Tina Skinner
Resource Acquisitions: Karl Bremer
Copy Editor: Dinah Roseberry
Photo Editor: Ginger Doyle
Book Designer: RoS

Special thanks to Concrete Stone & Tile Corporation and Hanover Architectural Products for technical advice, and to the many talented contractors and landscapers whose work is featured anonymously in this book. All of the images in this book were made available by VERSA-LOK Retaining Wall Systems and Willow Creek Concrete Products. For more information, visit www.willowcreekpavingstones.com or call 1-800-770-4525.

Introduction

What decks were to the 1990s, paver patios are to the twenty-first century. A paver patio has become the ultimate in outdoor living surfaces, and homeowners have been quick to discover the beauty, utility, durability, and ease of maintenance that comes with a properly-installed "interlocking" paver application.

Just as renowned architect Frank Lloyd Wright revolutionized the way we think about relationships between buildings and land, the myriad design possibilities of paving stones have redefined our perspective of landscape architecture to give your home and surroundings a sense of unity and place.

This book provides a dual purpose, providing both a portfolio of ideas for anyone hoping to beautify their landscape with concrete pavers, as well as practical, step-by-step instruction and useful tips should they undertake the installation themselves.

Moreover, this book offers up a peek at the future of paved surfaces with a permeable paver project. While the concept of permeable pavers may be relatively new now, it's a term you're going to be seeing more and more in the coming years as the world becomes increasingly sensitive to the planet's environmental needs.

Pavers 101

A Walkway to Remember

In theory, a walkway is small in scale and a logical first project for anyone undertaking their first step with paver installation. Following a short lesson in patterns, we'll show a sample walkway to illustrate how three different patterns are created. This primer should set you on your way to figuring out more complex designs, if you so desire. A gallery of walkways follow, and there will be plenty of additional inspiration throughout the book.

Laying the Groundwork

Professionals and do-it-yourselfers alike can install interlocking concrete pavers successfully. You probably will have to rent some tools that the average homeowner does not have. But if you take your time and pay particularly close attention to the base preparation, you will be pleased with the results.

For a simple lesson in various ways to lay standard pavers, we created this basketweave-pattern walkway. The pattern gallery that follows will help you choose the look that is right for you and your project. For more pathways, a gallery section at the end of the book will offer up a host of ideas.

1 A sidewalk project with a 90-degree turn gets its start in the standard way—with an excavation of 7-8 inches of topsoil to create a level, subsurface base, followed by 4 inches of crushed stone.

A Walkway to Remember

7

2 A vibrating-plate compactor goes to work.

3 Pipe "screed rails" are laid atop the crushed-stone base.

4 The base can be leveled by hand around the rails using a simple compacting tool.

5 A 1-inch sand layer is applied, first down one side, then the other, and then filled in the center.

6 The sand base is finished using a level both as screed and accuracy check.

A Walkway to Remember

7 The pipes are slid from the finished sand base to a new area on the stone base that needs a sand bed, and the level checked before finishing.

8 A trowel transfers fresh sand to the indent left by the pipe after it is removed.

9 Placing the first paver.

Tips for Success

Excavation

Before digging, always call your local utility company to locate any underground lines. In general terms, a minimum of 4" of compacted aggregate base is recommended for patios and walkways. Add 1" for the depth of the bedding sand and another 2-3/8" for the paver thickness to determine the total depth to excavate. Excavation should be 6" wider than the finished pavement's dimensions on sides where edge restraint is to be used. Slope and grade are important to ensure proper runoff. It is best to plan at least a 1/4"-per-foot drop, but try not to exceed 1/2" per foot.

Base Preparation

As with any building project, the finished pavers will be only as good as the construction of the base. For this reason, this is the *most* important part of the installation process.

Place and compact uniform layers of aggregate base material throughout the excavated area until the required depth and slope is reached. Best results will be achieved by using a vibrating-plate compactor. Slightly moisten dry base material and compact in layers of no more than 4 inches at a time. Start at the perimeter and work your way toward the center, overlapping each previous pass. Make at least two complete passes to create a flat smooth base.

After final compaction, check the entire area for proper pitch and level conditions. The base should now reflect the final grade of your pavers. If you were to place a straight edge on the surface, there should be no more than a 1/4" gap at any point along the straight edge.

Edge Restraints

The borders for your layout design may now be put into place. The edging is laid directly upon the crushed-aggregate base and secured with 10" steel spikes. One spike should be used every two (2) feet for walkways and patios and every one (1) foot for driveways and radii.

Sand Setting Bed

Lay the screed guides (1" outside diameter electrical conduit, strips of wood or other suitable rigid material) on top of the compacted base material 4' - 6' apart and parallel. For narrow areas such as walkways, the PVC edging can be used as a guide with a notched 2" x 4" board.

Installation of Pavers

Lay the first pavers against the longest edge of the project area such as a foundation, driveway or edge restraint. Start at the low end of the grade if possible and work uphill to keep pavers from shifting during construction. Select pavers from multiple cubes for optimum color blending. String lines will assist in assuring straight lines.

Compacting and Sweeping

A protective layer between the paving stones and the plate compactor during the final compaction will avoid scratching the surface of the pavers. The application recommended is the use of a mat, cardboard, thin plywood, carpeting, or soil separation fabric.

Basketweave Pattern

For a simple lesson in various ways to lay standard pavers, we created this basketweave-pattern walkway. The pattern gallery that follows will help you choose the look that is right for you and your project. For more pathways, a gallery section at the end of the book will offer up a host of ideas.

1 Border pavers are called the sailor course if laid end to end as in this example, or soldier course if laid side to side. A level provides a straight edge to test the alignment.

2 After creating the border, the first paver of the basketweave pattern is placed.

3 The basketweave pattern is continued.

Pavers 101

4 A basketweave pattern, before infilling with sand and final compaction.

Caring for Your Pavers

There are a variety of concrete cleaners in the marketplace. Contact your local dealer for their recommendation on which cleaner to use for your particular pavers.

Although it is not necessary, sealers can help enhance colors and resist stains. Sealers are not permanent and must be reapplied every 2-3 years. You must wait at least a full season before applying a sealer.

Herringbone Pattern

A herringbone pattern actually derives its name from a textile pattern, the result of a weaving process that left short parallel lines slanted in alternating directions. The pattern can be found on woven cloth dated back to 5000 years BC, and has evolved in applications from fabrics to masonry, as well as wood and tile parquetry.

After finishing the basketweave sample with a border, we begin a new pattern.

1 The first paving stone in the new section is set at a 45-degree angle. The herringbone section will not have a soldier course around its edges.

2 The first row continues with pavers placed parallel to the first one.

3 The second row is set, neatly fit into the recesses created by the first row.

4 The rows are continued to the end.

5 Use a straight edge to mark where the outer pavers must be cut.

A Walkway to Remember

6 The marked pavers are moved to the side, then fitted with new neighbors for the final row. These new neighbors are likewise marked for cutting.

7 The pavers are cut using a concrete saw.

Pavers 101

8 The trimmed pieces are reassembled on the sand base and fit like a snug puzzle.

What About Weeds?

Weeds will not grow up from below the pavers through a properly laid and compacted gravel and sand base. However, weeds and grass will result over time from airborne seeds. You may use polymeric sand between joints to prevent weed growth and ants, as well as to reduce washout.

PAVE LIKE A PRO: Measure the spaces for the offset pavers in that first row, making sure that they all fall on the same line. This way, your design stays even, and you'll only have to cut filler pieces of one size – in this case, half of a paver for each space.

Running Bond Pattern

As its name implies, this is a pattern that, once underway, runs off on its own and is fairly simple to apply.

1 A ruler or tape measure is an important tool for measuring the half-bond between the rows of pavers.

2 Ribs on the sides of the pavers maintain the proper spacing between rows.

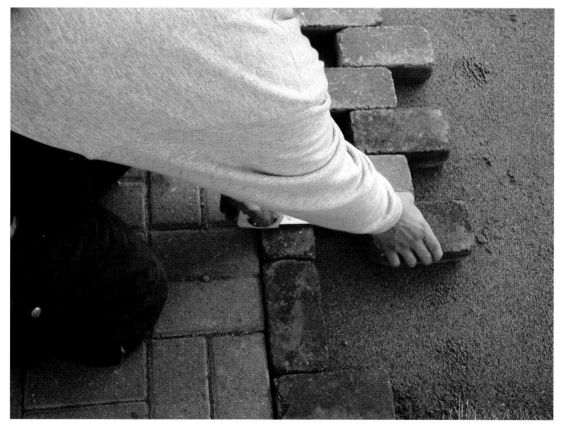

3 Once started, a straightedge keeps the pavers properly aligned.

Pavers 101

Finishing the Walkway

Whichever pattern you've chosen, the finishing process is the same.

1 Once the pavers are set on the sand base, more sand is shoveled on top.

2 The new sand is swept between the pavers to fill the cracks.

3 For stability, the walkway will be bordered with a plastic edge restraint system. Here the edging is measured for size …

A Walkway to Remember

4 ... and cut to fit.

5 Stakes secure the edge restraint.

6 A rubber mallet taps an errant paver back into place before compacting.

7 Finally, pass the vibrating-plate compactor over the pavers at least twice.

A Walkway to Remember

Gallery of Patterns

These sample grids should help you visualize—and describe to your landscaper or paver supplier—the look you are going for.

Running Bond
To see this pattern at work, see page number 107.

3-to-1 Running Bond

Double Basketweave

Pavers 101

Single Basketweave

Herringbone
To see this pattern at work, see page number 122.

Modified Herringbone

A Walkway to Remember

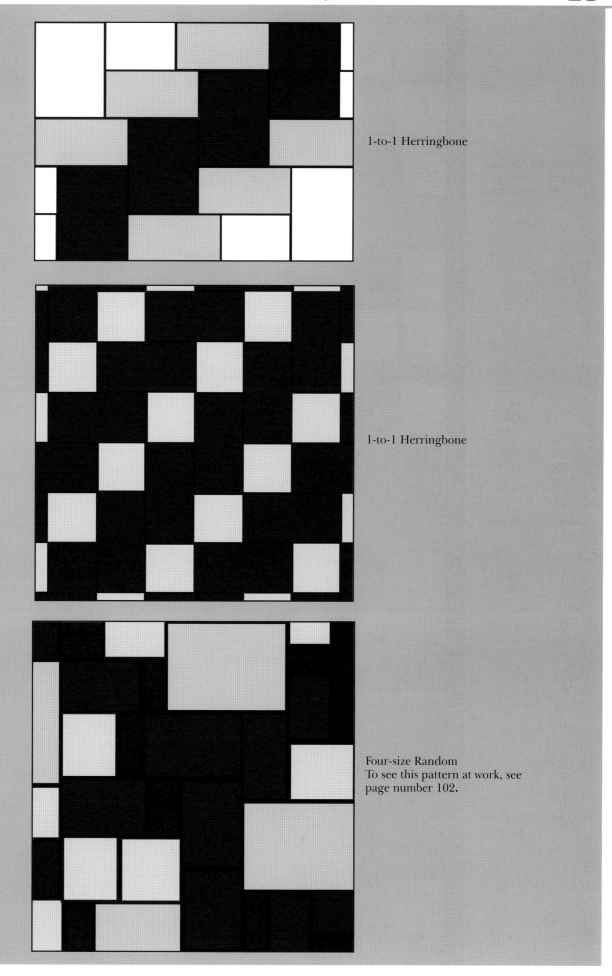

1-to-1 Herringbone

1-to-1 Herringbone

Four-size Random
To see this pattern at work, see page number 102.

24 | Pavers 101

Modified Herringbone
To see this pattern at work, see page number 102.

Modified Herringbone

Modified Herringbone

A Walkway to Remember

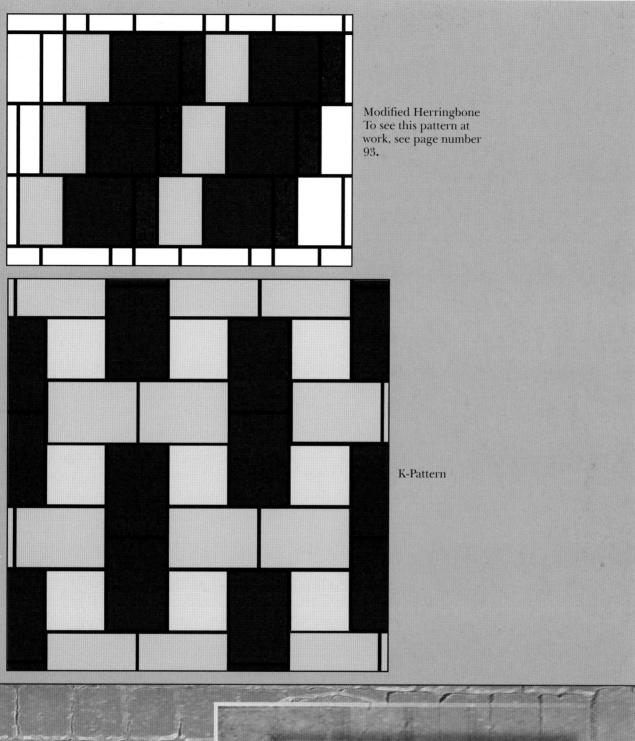

Modified Herringbone
To see this pattern at work, see page number 93.

K-Pattern

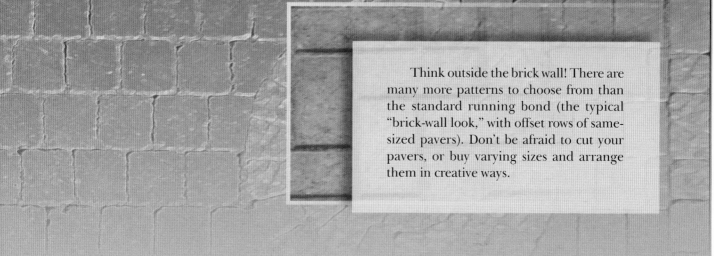

Think outside the brick wall! There are many more patterns to choose from than the standard running bond (the typical "brick-wall look," with offset rows of same-sized pavers). Don't be afraid to cut your pavers, or buy varying sizes and arrange them in creative ways.

Paver Patterns at Work

With the skills learned in this book, all these paver patterns are possible. The secret to success is properly planning and prepping the space, and then taking the time to fit, measure, mark, and cut the stones for a perfect fit.

A simple herringbone in two tones of pavers randomly assembled harkens back to historic brick patios.

A Walkway to Remember

Natural stone was set in concrete for the lead up to a patio with a grand, semicircular introduction.

Into this four-size herringbone pattern, a truly skilled paver set an eye-catching nautilus design.

A double border flanks a herringbone interior.

Two patterns meet at a thin swath of natural stones.

Four sizes of paver stones are combined for a random pattern set against a jewel-like border.

Small circle insets create added interest in this simple running-bond path.

A Walkway to Remember

This poignant change of color sets a target-like design into this red patio.

Patio Project with Fire Pit

Here's a wonderful project, conceivably achievable in a weekend's time with friends on board. The circle can be as big or as small as your ambitions. The *wow effect*, though, is immeasurable. This introductory chapter takes you step-by-step through the creation process of a back yard transformed, Cinderella-like, into a beautiful patio setting.

As shown, a family invested the sweat equity needed to create a wonderful backyard room with fire as its centerpiece.

Calculating Your Needs

BASE MATERIAL CALCULATIONS:

The following formula is used to calculate the amount of ¾" aggregate base material:

Step 1: Length x Width = Area (Measured in square feet)
Step 2: Area ÷ 12 x Inches of Base Required = Total Cubic Feet of Base
Step 3: Cubic Feet of Base x 1.25 (compaction factor) ÷ 27 = Yards of Base Needed

BEDDING SAND CALCULATIONS:

The following formula is used to calculate the amount of bedding sand needed (1 inch to 1½ inches depth):

Step 1: Length x Width = Area (Measured in square feet)
Step 2: Area ÷ 12 x 1" (typical thickness) = Cubic Feet of Sand
Step 3: Cubic Feet of Sand ÷ 27 = Yards of Sand Needed

Patio Project with Fire Pit

A backyard firepit was an instant hit and a favorite family hangout. Still, it needs panache, and a bit more durability given the regular traffic. New wooden steps descending from the patio will be taken into consideration as we plan the height and circumference of the patio, as well as a short walkway.

1 With the exception of a compacting machine and block splitter, we're going to do this primarily with elbow grease. Most of the tools needed were already at hand: shovels, metal rakes, and a broom.

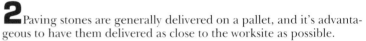

2 Paving stones are generally delivered on a pallet, and it's advantageous to have them delivered as close to the worksite as possible.

3 A block splitter can be rented to make the job easier, often from the same supplier that deals in paving stones.

4 Setting elevations is made easier with a laser transit. This also may be rented, but its use may be beyond the technical ability of some do-it-yourselfers.

5 A can of white spray paint was used to mark the patio perimeter right on the ground.

Patio Project with Fire Pit 37

6 The real work begins—excavating the site.

7 The curve of the trench begins to take shape. The excavated area should extend approximately 6 inches beyond the area you intend to pave; this provides an adequate foundation.

8 Excess soil and rocks are loaded into a trailer to haul away.

Patio Project with Fire Pit

9 A level subsoil surface is very important to prevent uneven settling. A plate compactor should be run over the excavated area.

10 The initial excavation is almost complete, and this overhead image illustrates the amount of soil that will need to find a new home. If you're removing nice topsoil, you might want to plan ahead for an area of the yard that needs to be filled in, or an attractive berm that can act as a focal point for plantings and provide visual and sound blockage for privacy.

11 The laser transit is put to use to determine the surface level upon which the pavers will be set. Stakes are set as placeholders to the height of the finished patio. We want a minimum of 2 inches of compacted gravel, in addition to the paver height. A very slight slope of ¼ inch per foot is calculated for the patio to allow for drainage.

Patio Project with Fire Pit

12 Gravel for the base is hauled and distributed throughout the area.

13 A vibrating-plate compactor is run over the excavated area to compress the gravel base. Each pass should overlap the previous one by about 4 inches. Start at the outer perimeter and work toward the center.

14 Using a compactor, compress the gravel base evenly. You should make at least two complete passes for each layer. Compaction should be performed in one direction, then a second time at a right angle to the first compaction.

15 Once compacted, be sure the gravel still meets the thickness you marked on the driven stakes.

Patio Project with Fire Pit

16 An area found to be low is filled with another layer of crushed stone, raked level, and then compacted with a manual tool. Likewise, the hand compacter can be used to level over the holes left when the stakes are removed.

17 The fire pit is fitted in place.

18 A 1-inch setting bed of concrete sand can now be spread on top of the compacted base.

19 The pipes can be dragged back and the leveling process continued.

20 A hand trowel is used to demarcate the edge and give a final finish to the tracks left by the pipes.

21 Beginning at the inner circle created by the fire pit, pavers are laid in a semi-circle working outward. A buddy system works well, allowing one person to hand off pavers, while the other focuses on placing them.

22 Spacing between pavers should not exceed 1/8". Set pavers lightly on the sand, never press or hammer them in. If you are doing the project over a couple of days, cover the entire area with plastic overnight if rain is expected.

23 A series of planks creates platforms for the installer to move across, which prevents the displacement of pavers with unequal weight.

Patio Project with Fire Pit

24 More buddies get involved as the work progresses, creating a series of stacks ready for the installer to set in the expanding circle.

25 A hand trowel finishes the edge, ready to meet the final row of pavers.

26 Small piles of sand are placed around the newly laid patio.

27 Spread and sweep sand over the entire top of the pavers using a stiff bristle broom.

28 Vibrating with a plate compactor will force the sand between the joints to stabilize and level the final surface. Excess sand should be swept into the joints. Re-sweep as necessary.

29 To provide stability and decorative embellishment for the fire circle, we will stack two layers of pavers around it and create a stone enclosure. Place beads of concrete adhesive on the innermost circle of pavers.

30 Stack the first layer of pavers.

31 More glue on top of the first stack will hold the second row of stones, which rises just to the lip of the fire circle.

32 The finished fire pit surround.

33 Now that the perimeter of the patio is clear, it's time to think about the stepping stones from the deck steps to the patio.

34 The process for creating a level surface is different here, because where the pavers were flat and even on the bottom, a natural-rock path will have crevices and inconsistencies. The stones are set in a sand and soil base, and tested for stability.

35 Sand is added and the surface groomed as needed for individual stones.

Pavers 101

36 The fit of the final stone is carefully checked.

37 The final step is spreading topsoil around the perimeter of the project and seeding it for grass.

Patio Project with Fire Pit

38 The finished patio, path and all, awaiting only water and sunshine for a perfect finish. In the background, a freestanding wall provides seating around the perimeter of the patio; keep reading for tips and techniques for building walls.

A Study in Similar Masterpieces

Because we studied a circular patio for our sample project, we start off with an inspiring study of patios in the round. A range of patios, from petite to party-scale, is offered up to help you envision a project perfect for your backyard. Most incorporate circles, though many offer a mix of other patterns that are perfectly compatible with the skills taught in this book.

Patio Project with Fire Pit

Circles of terra cotta-toned pavers emerge and descend into this expansive patio, taking full advantage of the property's graceful views. The "embossed" platform makes a great spot for a round patio table to echo the motif, while the "engraved" sitting area corrals chairs around a warm fire pit. The surface was created using a four-size herringbone pattern.

A round wall encloses three large urns for building impressive, Olympic-torch-like bonfires. A patio paved with permeable pavers creates the impression of ripples on water and reduces runoff from the lakeshore property.

Patio Project with Fire Pit

An expansive pattern of pavers circles from the center. The blocks were tumbled in the manufacturing process and created in a variety of tints, giving them the look of aged cobblestones.

A round patio is paired with a round outdoor table for a unified look. Loose stone in matching hues and small shrubs provide a wonderful transition to the lawn.

A soft border and stepped wall contain a complex pattern of intersecting circles on this patio.

Bold colors help define—and integrate—this large circular pattern in the patio and surrounding, matching stone garden beds.

Patio Project with Fire Pit

Circle patterns in the pavers naturally segregate the fire circle area from the dining table area.

A fish scale design swirls in reds, taupes and grays.

Patio Project with Fire Pit 61

Matching wall, columns, end caps, and paver stones unite.

This dramatic round patio is perfect for an inset fire circle. The heat is kept local by a half-moon wall, which also provides convenient seating.

Rejecting straight lines, a small patio gets big flair with a circular inset and contrasting border.

Right: **A freestanding wall** shelters a circular patio, poised midway on a descent to the lawn.

Patio Project with Fire Pit

A four-size random design fills a double circle border, an artful pausing point in the garden.

This tabletop firepit brings a luau atmosphere to this tropical patio.

Offshoots into the backyard house a fire circle, flagpole with walled enclosure, and an outdoor dining table; at the end, a final round pad leads to a private dock.

A circular design in these pavers creates a great pad for a round above-ground swimming pool, echoing its curves.

A grill turns a fire pit into a cook center.

An extensive patio illustrates how our starter project might be gradually expanded over time to conquer more of the back yard.

This fire circle and adjacent gazebo provide a place to relax after a round of golf.

A charming fire pit with grill top is ready for a cookout, set into pavers warmed by the sun.

Patio Project with Fire Pit 67

A trim of light stone draws your eye to this walled retreat under a giant oak. Wide columns make great gallery-style displays for Grecian planters. A fire circle waits inside.

Putting Up Walls

Our previous project included a short wall encircling part of the patio. These walls are wonderful ways to define a space, and they also add built-in seating that never blows away, breaks down, or needs painting. Here are easy steps for getting started. For more complete details on building segmental retaining walls, see *Building Outdoor Environments with Retaining Walls* and *Retaining Walls: A Building Guide and Design Gallery*, two comprehensive books available in fine stores everywhere!

1 Like a patio or any paver surface, a successful wall begins with excavation and compaction to create a uniform base. Stakes ensure the proper depth of base material.

PAVE LIKE A PRO: If you are working with mixed block sizes, remember to lay out your wall both on paper and in real life. Building a test wall on the ground—like a puzzle—will allow you to visualize your project and be sure you have all the blocks and sizes you need. In this stage, check color placement if you are using multiple shades, and be sure that your varying geometric shapes add up to a flat plane on top. Be sure to lay the wall down when doing the test-build, so you can pull stones right from the bottom row, gradually taking away layers to place in the actual wall. If you build the test wall standing up, you'll have to disassemble it to get to your first stones.

Putting Up Walls 69

2 Add a 6-inch layer of crushed stone.

3 Smooth using a rake.

4 The vibrating-plate compactor should be used to thoroughly compact your gravel.

5 At least two passes with the compactor, and the base is ready for the sand bed.

6 Two pipes (screed rails) are laid parallel and checked for level before adding the sand.

Putting Up Walls

7 Sand is leveled with a screed across the pipes, and the level double checked.

8 Starting with the corner column, the base units are set.

9 Place beads of concrete adhesive on the top of each unit.

10 The next course of retaining wall units is set to overlap the joint below.

Putting Up Walls 73

11 Keep stacking until your desired height is nearly reached; factor in the extra height that a stone cap on the column will add.

12 When the first corner is complete, begin a wall extending from it using the same stacking techniques.

13 Check the alignment of each course as you go along before placing adhesive for the next course.

14 The finished wall.

Walls to Remember

A close-up of a wall, color coordinated to match the paving stones.

Two simple freestanding walls set four courses high gracefully encircle a paver patio and terminate in small columns.

A freestanding wall cradles another project and provides seating around a circular firepit.

Putting Up Walls

This fire pit is framed by a curved-edge patio inset with a herringbone pattern.

A wall provides more seating when this gardener has company, as well as easy access to the lush vegetation beyond.

Permeable Driveway Project

Pavers 101

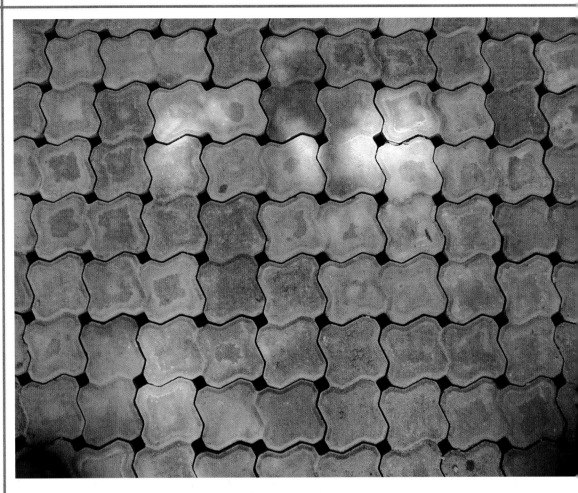

Permeable paving systems are the green landscaping solution you've been looking for to reduce stormwater runoff from your property and beautify its surface at the same time. An interlocking concrete paver system with an environmental twist, permeable pavers can create usable spaces that previously were unsuitable for paving—especially those now subject to new federal Clean Water Act stormwater regulations and in areas near protected waters.

The Aqua-Loc system used in this example permits stormwater to drain through aggregate-filled voids between the pavers into subsurface detention areas, where it's directed through a series of natural filtration processes before gradually exiting the system. By creating subsurface detaining and filtration areas, the need for costly, space-wasting detention ponds is dramatically reduced or even eliminated.

Beneath the pavers, Aqua-Loc relies on an innovative system comprising an aggregate bedding course, followed by a 4-inch filtration layer of drainage aggregate, an additional 12 inches of subgrade filtration rock, a perforated collector pipe, and an outlet pipe.

Permeable pavers are an environmentally sensitive alternative for projects such as driveways, parking areas, and patios in areas where stormwater runoff is undesirable or restricted.

Our sample project isn't a simple undertaking. It's an extensive driveway and parking area, and required a great deal of excavation. Moreover, slope and drainage were critical, and were done in consultation with professionals. Still, it's worth visiting for the do-it-yourselfer ready to undertake an extensive project, and helps illustrate how much labor is involved.

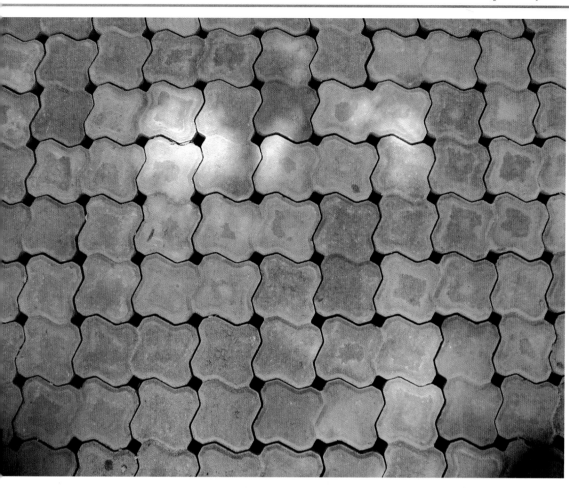

1 Permeable paving leaves gaps for easy infiltration of rainwater.

Edge Restraint Calculations:

The following formula is used to calculate the amount of edging needed:

Step 1: Measure the total linear footage of job, less any existing edge (foundation, curbing, wall etc.)
Step 2: Total Lineal Feet of Edging Required ÷ 8' = Total Pieces of Edging Required.
Step 3: Spikes Required: Driveway, 1 per 1 L/F; Walkway, 1 per 2 L/F

2 A Bobcat with a front-end loader is put to work excavating the site for a driveway and double garage bay entrance and parking area.

Permeable Driveway Project

3 Likewise, the machine is a great labor-saver for distributing crushed stone and doing the initial compacting and leveling. An 8-inch sub-base of 1.5"-3" aggregate is recommended for residential driveways.

4 Elbow grease and a rake are the only way to get into the corners and to get an even level for the crushed stone.

5 A vibrating-plate compactor should be passed over the gravel base at least twice.

Permeable Driveway Project

6 PVC pipe along the edge of the graded driveway will assist with drainage. Drainage is laid into the gravel.

7 The driveway must be graded away from the house for proper drainage.

8 Edge restraints will be used around the perimeter of the driveway and parking area, and in this case, are installed first.

11 The permeable pavers are placed first in a row along a straight edge of the driveway, facing the garage. We'll work outward from here.

Permeable Driveway Project

9 Flexible edge restraint will be used to create a semicircle around the entryway.

10 The border pavers are carefully placed next to the semicircular edge restraint.

12 The border pavers are removed to allow for placing and marking the permeable pavers with a compass, working back from the edge restraint.

13 The cut permeable pavers are fitted into their places next to the border stones.

14 The end of the driveway terminates in a flair at the street, and a soldier course follows the edge restraints.

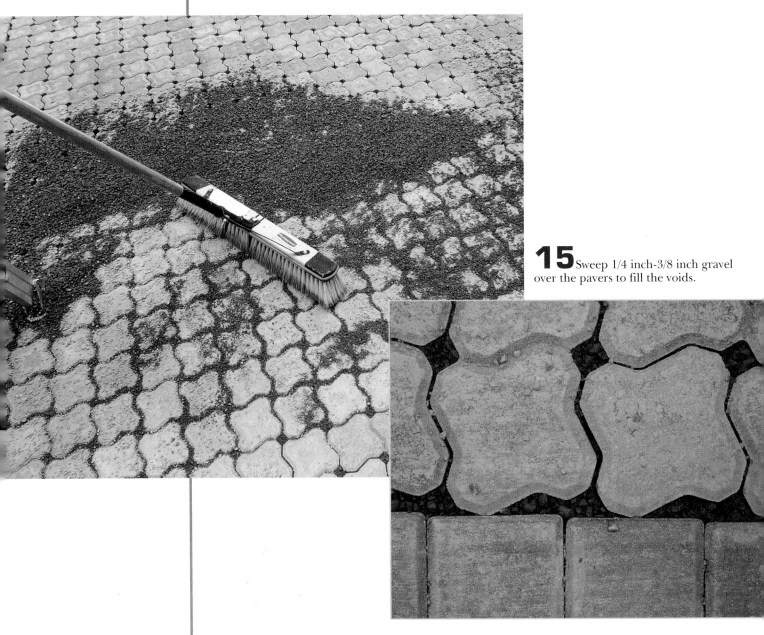

15 Sweep 1/4 inch-3/8 inch gravel over the pavers to fill the voids.

Permeable Driveway Project

16 Two passes with the compactor completes the installation.

17 A view from the street shows the finished project.

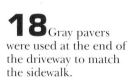

18 Gray pavers were used at the end of the driveway to match the sidewalk.

Gallery of Great Ideas

Pavers 101

A two-tone border corrals a random pattern in this handsome driveway that defines "curb appeal" for a new home.

Driveways

This simple and sophisticated driveway brings a touch of class to this residence.

This four-size herringbone design stretches from street to house in the spirit of a popular trend: the full paver driveway. Two free-standing columns light the way, one holding a house-number plaque. A central garden creates a useful and beautiful traffic loop.

Another great paver driveway features a large parking pad close to the house, sprawling out before a three-bay garage. It is the perfect ultradeep footprint for guest parking, a yard sale, or washing the cars.

Nothing is left to interpretation in this beautiful paver driveway; inlaid tan pavers define driving lanes and parking spaces so everyone is sure where to steer.

This parking pad and driveway move seamlessly into the front walkway, all in a white-gray scheme that provides appealing contrast to the ashlar stone facing on the home.

A traditional herringbone entry is fitting for a Tudor home straight out of merry old England.

Gallery of Great Ideas

Rich earthy hues interplay in a spiraling series of circles that provide a grand entry to this home.

Entryway Patios

A major trend in home landscaping has been the incorporation of small entry patios or courtyards in the front yards of homes. The concept has come full circle since the days of yesteryear, when front porches were commonplace and backyard decks and patios didn't exist. People are spending more time at home entertaining, and they want to utilize more than just the back yard for guests and socializing. A well-designed front-entry courtyard accomplishes that while also updating your home and adding value and curb appeal to it.

With a front-entry makeover, you'll have a place to welcome visitors and relax in the public arena of a peaceful community. In some instances, they may serve simply as a resting point for grocery bags in the process of transferring from car to pantry; in others, they act as crown jewels within the setting of a gardener's lush landscape.

Bullnose caps crown this short flight of steps to a circular landing at the entry to this home.

A circular design makes a visual transition around the corner of this home's front walkway.

Gallery of Great Ideas

Two round landings cap this front walkway—perfect for lingering moments as guests make their way in or out.

A round patio meets elegantly with a herringbone path that wanders off into the garden.

Slate-blue and brick-red tones in the entry walk pair beautifully with a home. The existing concrete stoop also was covered in pavers to complete the makeover.

Gallery of Great Ideas

A regal façade deserved a regal approach. An elegant front patio with rounded steps and landing raised on two courses of stacked retaining wall units provides a true sense of arrival. The low wall helps to contain and define the gently sloped patio area from the lower lawn.

A variety of patterns, heights, and widths make this short walk more entertaining.

There are no straight lines leading to this home, transforming the approach into a stroll.

Remarkable stone steps are flanked by Versa-Lok freestanding walls with inset lighting while below, a huge wheel of pavers completes a symmetrical façade.

An octagonal patio space keeps the eye moving. Natural stones form a makeshift path to the yard.

Lighting and potted plants provide safety features, highlighting edges and elevation changes.

Gallery of Great Ideas

These stairs were built using Versa-Lok retaining wall units, with retaining wall caps used for stair treads. Pavers matching the brick exterior topped it off.

This set of cascading circles eases the transition to crushed black stone with a trimming row of pavers that appears almost as a third step.

An elevated entryway patio acts as an inviting extension of the front door.

Gallery of Great Ideas   99

Pavers color coordinate with stone facing for an impressive entryway. Bullnose pavers finish off the stair treads nicely.

What's That White Film?

It's most likely efflorescence, a white deposit that sometimes appears on the surface of concrete, masonry or clay products. Efflorescence is often a natural by-product of cement hydration that appears as a white film when carried to the paver surface by moisture. Efflorescence is unsightly when it first appears but is not indicative of a flawed product. In many cases it is scoured off during installation but it can be removed using specialty cleaners. It will lessen and dissipate over time as the pavers naturally weather.

The front of this home is a visual delight, connecting the residents with their neighborhood in a series of gathering, play, and gardening areas.

A half wall adds a sense of security to an elevated entryway patio.

Gallery of Great Ideas

Perfect Paths

A winding ribbon of cobblestone pavers in a herringbone pattern and soldier course makes for an enchanting walkway.

A serpentine path leads from this paver drive in a matching design. A corner bed nestles a street lamp and punctuates a welcoming walk.

This four-size herringbone path in aged grays is paired with a red brick border, which transitions the path perfectly to garden beds sprinkled with red mulch.

A herringbone pattern set on an angle causes visitors to give this relatively short path a long look.

Gallery of Great Ideas

A path lined by low walls paves the way around a house.

Below left: **This simple herringbone pattern** winds around a garden, meeting concrete angels and other ornaments as it goes. A circular design spins around the corner turn.

Below right: **A natural stone path** links two sections of pavers in circular and basketweave designs. River rock holds each magnificent stepper in place.

An elegant basketweave pattern.

A herringbone pattern marches past bushy hosta plants.

Coordinating mulch makes this path unobtrusive and natural.

Gallery of Great Ideas

This path leads past walls, benches and free-standing columns nestled into the flowerbeds.

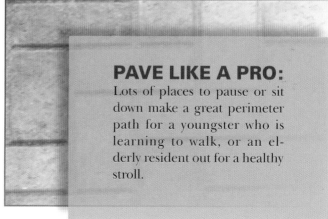

PAVE LIKE A PRO: Lots of places to pause or sit down make a great perimeter path for a youngster who is learning to walk, or an elderly resident out for a healthy stroll.

Where this path leads is hard to tell—but if the scenery is anything like this section's, you'll definitely want to keep walking.

At the entrance to this building, a geometric path helps reinforce a stately façade.

The circular pattern at the turn in the walkway subtly steers approaching guests toward the front door.

Two lanes are partitioned by a charming garden with a hummingbird feeder.

Gallery of Great Ideas

A brick staircase with multiple tiers is smartly dressed in red with sand-colored border wide enough to accommodate potted flowers. Wrought iron provides an important safety rail on the opposite side.

PAVE LIKE A PRO: If you're a fan of plants and flowers in decorative pots, rather than garden beds, plan your path accordingly with a double- or triple-wide border, so that your pots don't intrude on your room to walk. You'll be amazed at how this tiny detail makes a huge difference in the functionality of your path.

Pavers 101

A paver path cuts its way across a concrete driveway.

Circular insets turn this paved passageway behind the home into a place of respite.

Gallery of Great Ideas

Wonderful Walls

This outdoor bar and grill are constructed on retaining wall blocks capped with wide, counter-like slabs. A curved freestanding wall defines the outdoor room, and makes a great place to sit with a drink.

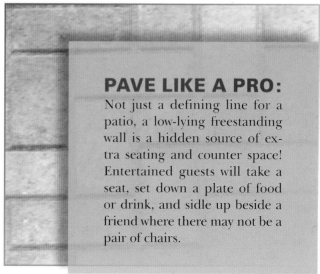

PAVE LIKE A PRO: Not just a defining line for a patio, a low-lying freestanding wall is a hidden source of extra seating and counter space! Entertained guests will take a seat, set down a plate of food or drink, and sidle up beside a friend where there may not be a pair of chairs.

A rounded wall helps provide patio dwellers, sunbathers, or readers a bit of privacy from the street below.

This wall features inset lighting to illuminate the front walkway, while two smart lanterns appoint the end columns.

Retaining wall products serve as great building blocks for a raised flower garden.

Gallery of Great Ideas   *111*

Mixing and matching walls makes for a visually interesting back porch on this sloped property. While tiered walls frame the multilevel patio against the house, an iron rail protects the drop-off without obstructing the view.

A series of walls served to tame a sloped front yard, providing level sitting places and terraced gardens.

These walls have been used to carve out a paved sitting area from a veritable backyard jungle of lush greenery. They line the open space, as well as every approaching and hidden path. A four-size herringbone patio lies underfoot.

A gorgeous round raised patio with a perimeter wall around the outside edge was the solution for this home on a slope.

This raised-bed garden provides a beautiful border for a paved driveway.

A freestanding wall provides a sunny seat as well as a patio border.

Gallery of Great Ideas 115

The Next Step

A set of steps is tucked into a tiered wall, which holds up an outdoor dining area and grill.

These gray steps use red pavers on stair treads for a stylish version of a painted caution line.

PAVE LIKE A PRO: As dramatic as a set of lone, freestanding stairs looks, take into consideration your family members, cold-weather conditions, and traffic patterns to determine if a no-railings staircase is really a good idea for your home.

Extra wide and deep steps allow for doorstep potted plants that welcome visitors.

Another eye-catching strategy: Light treads and dark risers help the steps stand out.

Rounded steps in mottled gray lead majestically from a spacious paved patio to a sliding glass door.

A pair of staircases provides dual access to this patio from different points in the yard.

A series of stairs and retaining walls helps navigate a sloped terrain.

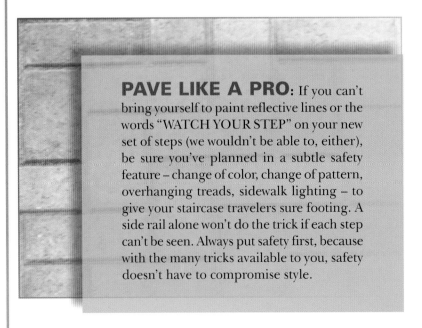

PAVE LIKE A PRO: If you can't bring yourself to paint reflective lines or the words "WATCH YOUR STEP" on your new set of steps (we wouldn't be able to, either), be sure you've planned in a subtle safety feature – change of color, change of pattern, overhanging treads, sidewalk lighting – to give your staircase travelers sure footing. A side rail alone won't do the trick if each step can't be seen. Always put safety first, because with the many tricks available to you, safety doesn't have to compromise style.

Gallery of Great Ideas

Round stairs and a landing are framed by pillars.

Offset brickwork helps to distinguish each step, while the integrated stairs serve as a retaining wall.

More Pretty Patios

This enormous patio made way for a tree and small garden, which will amply repay the effort with shade every summer.

A newly installed paver patio is nestled in fresh mulch around its free-form border.

This patio with a young garden gives outdoor entertaining space to a modest-sized home.

Gallery of Great Ideas 121

Sunbathers can enjoy this private retreat; in the evenings, the torches can be lit for an island-inspired gathering.

A raised patio built on top of Versa-Lok retaining walls turned this steep slope into an outdoor room with a lake view and access from above and below.

Winterizing a Paver Application

Concrete pavers are high-density units that resist deterioration from de-icing salts better than asphalt, ordinary poured-in-place concrete, and stamped concrete. The most recommended salt for ice removal is calcium chloride.

You can remove snow from a paver surface just like any other pavements: You can plow, shovel, or use a snow blower with confidence.

Paver applications and segmental retaining walls do not need concrete footings below the frost line. The flexible nature of the applications allows them to accommodate minor earth movement without damage.

Pavers 101

These pavers provide a great contrast to the shimmering blue water of an in-ground pool, and provide an extra wide path in a herringbone pattern.

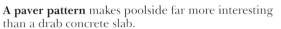

A paver pattern makes poolside far more interesting than a drab concrete slab.

For pools with tile designs—real, or faux—on their linings, bullnose pavers make great coping around the edge. Note the detailed corner work on the pavers.

This square fire pit, framed by a geometric design, is a refreshing variation from the typical circle. An ultrawide edge leaves plenty of room where warming plates or marshmallow sticks can rest.

Gallery of Great Ideas

A fire circle boasts seating built into free-standing perimeter walls.

Quarter-circle steps and landing are a great solution for a corner door. A four-size herringbone patio rests below.

Steps help unite two very different outdoor surfaces: white decking and slate-gray pavers.

This steep, unusable slope was transformed to a beautiful outdoor entertainment area by tiered retaining walls supporting a multilevel patio.

Gallery of Great Ideas

125

A set of stairs built with Versa-Lok retaining wall units provides an escape to a lower tier, where a curious green-thumb can walk among the plants. Above, the view from the dining table to the lower patio levels is impressive.

Approaching this home is like entering old-world Italy, with its archways and stone details sheltering the exposed loggia area. A paver patio in light tones is embellished with walls and covered corridors of burnt orange, slate gray and white stone.

Two tables hide beneath umbrellas, sheltered by a freestanding wall.

This lower-level patio is perfect for the family reunion, graduation party, or Independence Day cookout. With an overhanging deck *and* exposed space, all-weather parties can gather here. An exterior wall features inset spot lighting. A staircase in the foreground leads right to the kitchen for easy food access, while a grill pad in the background keeps the smoke of the grill away from conversing guests.

Gallery of Great Ideas

This incredible patio retreat features a four-size herringbone design underfoot in soft gray pavers, with a sitting area that is marked by an overhead trellis and white columns. Inside, chairs are placed near an outdoor stove—a less smoky, lower-maintenance fire alternative. The entire patio is raised above the lawn by a serpentine retaining wall and accessed by blue-gray concrete steps.